Guidelines for Reporting Occupation and Industry on Death Certificates

Authors
This document was written by Cynthia Robinson, Pam Schumacher,
Marie Haring Sweeney and Jose Lainez

Department of Health and Human Services
Centers for Disease Control and Prevention
National Institute for Occupational Safety and Health

4676 Columbia Parkway, Cincinnati, Ohio 45226

This document is in the public domain and may be freely copied or reprinted.

Disclaimer

Mention of any company or product does not constitute endorsement by the National Institute for Occupational Safety and Health (NIOSH). In addition, citations to websites external to NIOSH do not constitute NIOSH endorsement of the sponsoring organizations or their programs or products. Furthermore, NIOSH is not responsible for the content of these websites. All websites referenced in this document were accessible as of the publication date.

Ordering Information

To receive documents or other information about occupational safety and health topics, contact NIOSH at
Telephone: 1-800-CDC-INFO (1-800-232-4636)
TTY: 1-888-232-6348
E-mail: cdcinfo@cdc.gov
or visit the NIOSH website at www.cdc.gov/niosh.

For a monthly update on news at NIOSH, subscribe to NIOSH eNews by visiting www.cdc.gov/niosh/eNews.

For more information on this topic and to access a pdf version of this document, visit the NIOSH Industry and Occupation Coding and Support web page (http://www.cdc.gov/niosh/topics/coding/).

DHHS (NIOSH) Publication No. 2012-149
June 2012

Acknowledgements

NIOSH recognizes funeral directors, the National Association for Public Health Statistics and Information Systems (NAPHSIS), the National Center for Health Statistics (NCHS), and state health departments for their critical role in producing accurate and complete statements of occupation and industry.

We would like to thank the National Funeral Directors Association (NFDA), the National Center for Health Statistics (NCHS), and the Cincinnati College of Mortuary Science for their consultation in the development of this document.

Contents

Disclaimer ii

Ordering Information ii

Acknowledgements iii

Chapter 1: Introduction 1

        Importance of Collecting Occupation and Industry 2

        Note to Funeral Directors 2

Chapter 2: Completing Electronic and Paper Death Certificates 3

Chapter 3: Rules for Reporting Usual Occupation and Kind of Business or Industry 5

        Basic Rules 5

        Rules for Special Situations 5

Chapter 4: Completing the Occupation Item 9

        A) Instructions 9

        B) Important points to remember 10

        C) Generic occupations often reported that may require clarification 11

        D) Commonly confused occupations that require clarification 13

Chapter 5: Completing the Kind of Business or Industry Item 15

        A) Instructions 15

        B) Special Situations 17

Chapter 6: Scenarios with Several Jobs over a Lifetime 19

Chapter 7: Checklist: Reviewing Your Entries 21

Appendix A: Examples of Occupation that Need more Detail 23

Appendix B: Examples of Industry Entries that Need more Detail 29

Appendix C: Examples of Acceptable Entries for Both Occupation and Industry 35

Appendix D: The Vital Statistics Registration System in the United States 37

        Using Death Data for Public Health Action 38

Chapter 1
Introduction

By reporting accurate data on industry and usual or lifetime occupation of decedents, funeral directors and those involved in the registration process are helping to improve statistics on occupational mortality and worker health. This document updates the guidelines written in 1988 by the National Center for Health Statistics (NCHS) (DHHS Publication No. 88-1149). It is designed to help funeral directors complete the Decedent's Usual Occupation and Kind of Business/Industry items on electronic and paper death certificates. The National Institute for Occupational Safety and Health (NIOSH) reviews the quality of the occupation and industry reported, combines it with the NCHS mortality data, and reports U.S. occupational mortality trends.

The average U.S. worker spends a substantial part of his or her life at work. Workplaces or jobs may expose workers to risks or hazards that can contribute to injury or disease. Information about an individual's job and type of business provides a snapshot of the hazards he or she might have encountered on the job. Disease may appear immediately or many years after the worker left the job. The data on the occupation and industry of workers can point to health risks to which the workers may have been exposed.

In the United States there are approximately 5,000 traumatic work-related fatalities and tens of thousands of work-related deaths from illnesses reported each year. It has been known for many decades that exposures to hazards in the work environment can cause severe injury or illness and death in workers.

It is important to collect high quality, accurate, and complete responses on the death certificate for all decedents over age 14 that were ever employed, unemployed, or retired. Such data can be used for making decisions on creating safer and healthier work environments.

Importance of Collecting Occupation and Industry

Public health program planners and researchers rely on funeral directors to record the best information possible on potential risk factors, including potentially hazardous jobs held by decedents during their working lives.

The usual (longest-held) occupation and kind of business or industry of workers can reveal the national illness and injury burden by industry and occupation. Such information can also be used to help discover jobs that may have a high risk for death due to injury, cancer, or other diseases and for which prevention efforts can be concentrated or targeted.

Note to Funeral Directors

We understand that funeral directors are often limited by the amount and specificity of information received from survivors, but we believe that the enclosed tips will be helpful. The examples described as "adequate" are always preferred over those entries described as "inadequate," but "inadequate" data entries are acceptable when no other information is available. In this booklet we use the term inadequate to describe an entry that does not provide enough information for accurate coding of an industry or occupation according to standard classification systems.

Chapter 2

Completing Electronic and Paper Death Certificates

Who Provides the Industry and Occupation?

The individual who provides information about the decedent should be someone who knows firsthand about the decedent's job history. This person, hereinafter referred to as the informant, should be able to describe accurately the types of jobs held by the decedent during his or her life.

The informant is usually the surviving spouse or a first-degree relative (parents, siblings, children). Other family members or friends may be able to provide details about the decedent's longest held or usual occupation and kind of business or industry.

Getting the Best information

If the decedent had many different occupations and different places of business, it may be necessary to ask additional questions to determine the usual occupation and industry, such as:
- In which job did the decedent work the longest?
- What kind of work did the decedent do for the most years of his/her life?
- If the decedent had several jobs over a lifetime, in which job did he/she work the longest?

Chapter 3

Rules for Reporting Usual Occupation and Kind of Business or Industry

> **Remember**
> There will always be a corresponding *Kind of Business/Industry* for every *Usual Occupation* entry.

This chapter reviews the rules for reporting *usual occupation* and *kind of business or industry* for individuals who were ever employed during their life. However, many decedents may not have been in the workforce at the time of death; thus, there might not be an occupation response to report. The following rules provide guidance on what should be reported in such situations, and for responses such as child, homemaker, or retired, among others.

Basic Rules

1. If the decedent was ever employed, always complete the Usual Occupation and Kind of Business/Industry items.

2. Do not leave any blank occupation and business/industry items.

Rules for Special Situations

1. If the decedent is under 14 years of age, enter "*infant*" or "*child*" in the occupation or business/industry item.

2. *Student* (14 years or older):

- *Full-time student* with no employment:
  - Enter "Student" for Usual Occupation.
  - Enter type of school, such as "High school" or "College," for Kind of Business/Industry.
- *Part-time student* with employment (part-time or full time):
  - Enter his/her usual occupation and the corresponding business/industry.

Chapter 3: Rules for Reporting Usual Occupation and Kind of Business or Industry

3. *Retired:* **Never enter "retired."** If the decedent was retired, enter the kind of work done during most of his/her working life.

4. *Unemployed:* **Never enter "unemployed."** Enter the usual occupation and business/industry of the decedent if he/she were ever employed. If the person never worked, use rule # 8 (*Never worked*).

5. *Unknown:* Should be entered only after every effort has been made to determine the usual occupation and business/industry. *Unknown* should be your last resort, and must be entered on both *Usual Occupation* and *Kind of Business/Industry*.

6. *Self-employed:* If the person was self-employed, enter the kind of work performed, including a trade or craft, for the *Usual Occupation*. For *Kind of Business/Industry*, enter the usual occupation with the term "self-employed".

7. *Disabled* or *Institutionalized:* Enter the usual occupation and business/industry of the decedent if he/she were ever employed. If the disabled or institutionalized decedent never worked, apply rule # 8 (*Never Worked*).

8. *Never worked:* If the decedent was not a student or homemaker/housewife and had never worked during his/her life, enter "Never worked" for *Usual Occupation* and *Kind of Business/Industry*.

Example of Self-Employed

Usual Occupation	Kind of Business/Industry
Shoe repairman	Self-employed shoe repairman
Repairman	Self-employed Repairman
Auto Painter	Self-employed Auto Painter

Chapter 3: Rules for Reporting Usual Occupation and Kind of Business or Industry

9. *Homemaker*[1]: To determine the entry for the occupation reported as "homemaker," it is important to find out if the decedent:

    a. Worked for pay outside the home.
    b. Worked at his/her own home or someone else's home.

Homemaker vs. Domestic Occupation Flow Chart

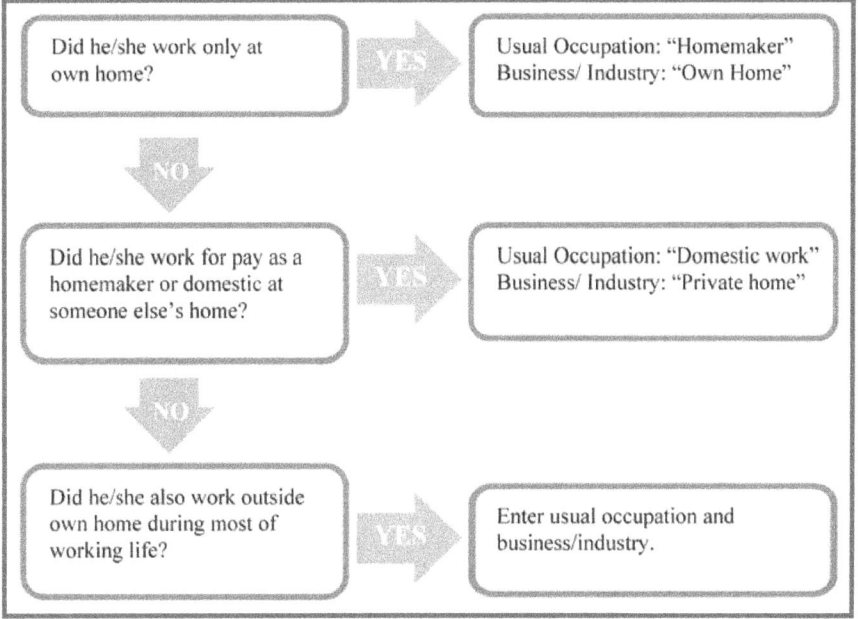

---

[1] Also known as "Stay-at-home mom," "Stay-at-home dad," "Stay-at-home spouse," or "Housewife."

Chapter 4

Completing the Occupation Item

> **Definition**
> **Decedent's Usual Occupation:** Type of job the individual performed during most of his or her life, or the job held for the longest time.

A) Instructions

In the space on the death certificate for *Decedent's Usual Occupation*, enter the word or words which most clearly describe the kind of work or type of duties performed by the deceased person during most of his/her working life.

The following information can help to determine the decedent's usual occupation.
- The usual occupation may be obtained by asking the informant to identify the longest held job. This applies to all individuals who, at the time of death, were ever employed, unemployed, or retired during his/her lifetime.
    - If the decedent had several jobs, identify the one that was held for the greatest number of working years.
    - It is important to note that this is not necessarily the last occupation of the decedent.
    - The usual occupation may not be the highest paid job or the most prestigious job.

Enter the longest held job onto the death certificate. The best response must be specific and descriptive. It should reflect job tasks and duties.
The following questions could be used to obtain more specific job tasks and duties:
  "What kind of work was he or she doing?"
  "What were his or her specific duties and tasks?"
  "How long had he or she been doing this kind of work?"
If the Decedent's Usual Occupation is not clear enough from the information provided by the informant, enter a few descriptive words. See examples in Table 1.

Example of Usual Occupation
A woman worked in the kitchen of a fast-food restaurant for five years. Later, she worked as a secretary for a car dealership for 25 years. Then, she worked as a care assistant at a retirement home for 10 years.

What is her *usual occupation and kind of business/industry?*
  Usual occupation: "Secretary"
  Kind of business/industry: "Car dealership"

Although she had three long-term jobs, her job as a secretary at the car dealership was the longest held (for 25 years). *Car dealership* is the type of business in which she worked as a secretary.

Chapter 4: Completing the Occupation Item

B) Important points to remember

It is important that the entry for occupation be very specific.[2] The following examples may be helpful when clarifying vague or unusual occupations:

1. *Apprentice or helper.* Enter if the person was an apprentice, trainee, or helper, in his/her occupation.

Examples

- Accountant's helper
- Electrician's helper
- Auto mechanic apprentice
- Plumber apprentice

2. *Business owner.* Enter if the decedent was the owner of a business. The usual occupation for the business owner is not necessarily related to the type of business.

Examples

Too General	More Specific
• Baker	• Bakery owner
• Printer	• Publishing firm owner
• Brewer	• Brewery owner

3. *Unusual occupations.* Enter the occupation given by the informant even if the occupation may sound unusual.

Examples

- "Sand hog": works in construction of underwater tunnels.
- "Printer's devil": refers to an apprentice printer.
- "Flatcar whacker": maintains and repairs logging rail cars.
- "Bed rubber": operates machine which smoothes stone blocks.

Table 1
The following example shows selected general occupations; however, *the more specific the occupation, the better the entry.*

General	More Specific
Salesperson	Insurance sales, advertising sales …
Assistant	Office assistant, dental assistant…
Manager	Grocery store manager, office manager…
Teacher	Preschool teacher, high school teacher…
Farmworker	Fruit farmworker, dairy machine operator, farmworker, livestock handler...

[2] refer to Appendix A for more examples of occupations that require further detail.

C) Generic occupations often reported that may require clarification

There are many occupations that describe a general type of work such as contractor, consultant, assembler, and technician. It is important to identify *the specific nature of the work performed* and the decedents' kind of business/industry.

1. *Contractor.* Enter the type of contract work the decedent did.

   Examples

   - Farm contractor
   - Plumbing trade contractor
   - Administrative contractor
   - Defense contractor

2. *Consultant.* For what are they providing consultation? The term "consultant" should include a specific occupation.

   Examples

   - Accounting consultant
   - Pharmacy consultant
   - Computer consultant
   - Nurse consultant
   - Dietitian consultant
   - Wig sales consultant

3. *Assembler.* Include the type of product that is being assembled.

   Examples

   - Toy assembler
   - Truck engine assembler
   - Furniture assembler
   - Household appliance assembler
   - Electronic assembler
   - Concrete buildings assembler

4. *Technician.* Include the specific technical process involved in the decedent's work.

Examples
- Medical laboratory technician
- Automotive technician
- Broadcasting technician
- Computer software technician
- Pharmacy technician
- Refrigeration technician

5. *Laborer.* Include the type of work performed by the laborer.

Examples
- Warehouse laborer
- Construction laborer
- Farm laborer
- Cleaning laborer
- Oil field laborer
- Landscape laborer

D) Commonly confused occupations that require clarification:

Some occupations may sound the same or may be difficult to discern. Ask additional questions to make sure that the correct occupation is entered. Below are examples of occupations that can be confused:

1. "Machinist," "Mechanic," "Machine operator":

- "Machinist": constructs metal parts, tools, and machines through the use of blueprints, machine and hand tools, and precise measuring instruments.

- "Mechanic": inspects, services, and repairs machinery.

- "Machine operator": operates a factory machine; for example, a drill press or winder.

2. "Homemaker," "Housekeeper," / "Domestic help" / "Housemaid":

- "Homemaker": manages the household while his/her spouse earns the family income. It is also known as "stay-at-home mom," "stay-at-home dad," "stay-at-home spouse," or "housewife."

- "Housekeeper" / "Domestic help"/ "Housemaid": performs household work, for pay, in private residences (not in own home). This is not the same as "cleaning service."

Refer to Chapter 3, item 9 (Homemaker) for more information.

Chapter 5

Completing the Kind of Business or Industry Item

> **Definition**
> **Kind of Business/Industry:** Type of business or industry where the decedent worked in his/her usual occupation.

A) Instructions:

In the space on the death certificate for Kind of Business/Industry, enter the type of business or industry that the decedent worked in during most of his or her life, or for the longest time. Do not enter company names.

1. Always enter the Kind of Business/Industry. It should be reported even if the Decedent's Usual Occupation is unknown.

2. Only use terms that clearly describe the kind of business/industry at the location where the decedent was employed.

- The terms should indicate both a general and a specific function for the employer. Examples of a specific function include livestock farm, automobile manufacturer, wholesale grocery, retail bookstore, road construction, and shoe repair service.

- The words "agriculture," "manufacturing," "wholesale," "retail," "construction," and "repair services" refer to general industry categories and therefore only describe broad or general functions. Combining general terms with a specific function will usually provide the best detail. See Table 2 for examples of combinations of business/industry and specific function.

Table 2	
General Industry	**Business/Industry with Specific Function**
Agriculture	Livestock farm, Fruit production
Manufacturing	Automobile manufacturer, Food manufacturer
Wholesale	Wholesale grocery, Plumbing supplies wholesale
Retail	Retail bookstore, Furniture retailer
Construction	Road construction, Building construction
Repair Service	Shoe repair service, Transmission repair service

3. To obtain more specific business/industry functions, ask the following types of questions:

- If the informant reports that the decedent worked in the healthcare industry, ask, "What type of location, hospital, doctor's office, or clinic?"

- If the informant reports that the decedent worked in the automobile industry, ask, "Did they manufacture, repair or sell automobiles?"

- If the informant reports that the decedent worked at a service business, ask, "What type of service did they perform?" Examples could be refrigerator repair, laundry service, cleaning service.

4. It is important to distinguish among manufacturing, wholesale, retail, and service companies. Even though a manufacturing plant sells its products in large lots to other manufacturers, wholesalers, or retailers, report it as a manufacturing company. Use the following as guidance:

- A wholesale establishment buys, rather than makes, products in large quantities for resale to retailers, industrial users, or to other wholesalers.

- A retailer sells primarily to individual customers or users but seldom makes products.

- Establishments that are engaged in providing services to individuals and to organizations include hotels, laundries, cleaning, advertising agencies, and automobile repair shops.

B) Special situations:

The following are important points to remember when entering Kind of Business/Industry:

1. Companies often operate more than one business or industrial activity, sometimes in the same location.

- Enter the business or industrial activity in which the person actually worked, not the company name.

Example

A chemist worked in a papermill operated by the Eastman Kodak Company.

Kind of Business/Industry should be "Papermill," not "Eastman Kodak Company" or "Camera factory."

2. Government organizations.

- Enter the level of government; for example, federal, state, county, or city.

- Enter the activity of the government organization and the name of the specific organization if it is responsible for several activities.

Example

Level of Government	General Activity	Specific Activity
City	Department of Public Works	- city street repair - city garbage collection - city sewage disposal

The best industry entries would be *city street repair, city garbage collection, or city sewage disposal.*

- Enter the name of the government organization only when its activity is clear.

Example

- U.S. Bureau of the Census
- City Fire Department

3. Businesses in own home.

- Some people conduct business in their homes. These businesses should be reported in the same manner as regular business establishments. Examples include a dressmaking shop, lending library, cabinetmaking shop, radio repair shop, and physician's office.

4. People who work in jobs at different locations.

- Some workers are required to work at different places rather than in a specific store, factory, or office. This includes people who normally work at various locations at different times. In such cases, report the kind of organization or industry for which they worked.

Example

Occupations	Kind of Business or Industry
Census interviewers	U.S. Bureau of Census
Building painters	Construction
Landscapers	Lawn service and landscaping

5. Domestic and other private household workers.

- If the name of an individual is given as the name of the employer, ask whether the person worked at a place of business or in a private home. For example, if the person was a domestic worker in the private home of John Q. Public, the correct entry would be "Private home." Conversely, if a person cleaned offices in a private home, such as a doctor or lawyer's office, the correct entry for Kind of Business/Industry would be "Doctor's office" or "Lawyer's office."

For some industries, common titles are not adequate. Refer to Appendix B for more examples.

Chapter 6

Scenarios with Several Jobs over a Lifetime

Scenario: 35 Years as a Welder and 5 Years as a Night Watchman

A man worked for 35 years as a welder in a steel-fabricating shop. Before retiring, he worked as a night watchman for 5 years.

What is his *usual occupation* and kind of *business/industry*?
   Usual occupation: "Welder"
   Kind of business/industry: "Steel fabrication"

Even though he was a night watchman for many years before retirement, he worked as a welder for a longer period of time (35 years). *Steel fabrication* is the kind of industry in which he worked as a welder.

Scenario: 27 Years as a Coal Miner and 8 Years Post-Retirement as a Custodian

A man worked for 27 years as a coal miner. After retirement, he worked for eight years as a custodian for a public middle school.

  What is his usual occupation and kind of business/industry?
   Usual occupation: "Coal miner"
   Kind of business/industry: "Coal mining"

His work as a miner accounted for the majority of his working years; therefore miner is his longest held job (27 years). Coal mining is the specific type of industry in which he was a miner.

Scenario: 22 Years as a Sales Clerk and 18 Years as a Homemaker

A woman worked for 22 years as a sales clerk at Macy's. Later she quit that job and became a homemaker. She was a homemaker for 18 years.

  What is her usual occupation and kind of business/industry?
   Usual occupation: "Sales clerk"
   Kind of business/industry: "Department store, retail"

Her work as a sales clerk accounted for the majority of her working years; therefore sales clerk is her longest held job (22 years). Department store, retail is the specific type of business in which she was a sales clerk. Notice that the name of the company, Macy's, is not a useful entry.

Chapter 7

Checklist: Reviewing Your Entries

Follow this checklist to improve the quality of the entries for *Decedent's Usual Occupation and Kind of Business/Industry*.

1. **Was the decedent 14 years of age or older?**

   ☐ Yes: Enter *Decedent's Usual Occupation and Kind of Business/Industry*.

   ☐ No: Enter "infant," "child," or "student" as it applies.

2. **Was the decedent retired?**

   ☐ Yes: Even if the decedent had retired, enter *Usual Occupation and Kind of Business/Industry*.

   ☐ No: Enter *Usual Occupation and Kind of Business/Industry*.

3. **Was the decedent self-employed?**

   ☐ Yes: Enter the kind of work performed, including a trade or craft, for the *Usual Occupation*. For *Kind of Business/Industry*, enter same usual occupation with the term "self-employed."

   ☐ No: Enter *Usual Occupation* and *Kind of Business/Industry*.

4. **Was the decedent a homemaker who worked outside own home?**

   ☐ Yes: Enter usual occupation and corresponding business/industry of the decedent if he/she worked outside own home during most of working life.
   Enter "Domestic work" for *Decedent's Usual Occupation* and "Private home" for *Kind of Business/Industry* if the decedent worked for pay at someone else's home.

   ☐ No: Enter "Homemaker" for *Decedent's Usual Occupation* and "Own home" for *Kind of Business/Industry* if the decedent worked only in his/her own home.

5. **Did you leave any blank entries?**

☐ Yes: Do not leave blank entries. Always enter *Usual Occupation* and *Kind of Business/Industry*. If the decedent never worked during his/her life, enter "Never worked" for *Usual Occupation* and *Kind of Business/Industry*.

☐ No: Good job! Both entries, *Usual Occupation* and *Kind of Business/Industry*, should always have a response.

Appendix A

Examples of Occupation Entries that Need more Detail

The following list consists of examples of occupations which require more detail. Included in this listing are examples of *inadequate* entries — those considered too general or needing more detail — as well as the *adequate* entries. The most frequent inadequately reported occupations appear in bold. Note that examples listed as adequate do not include all acceptable occupation titles. Refer to Appendix C for additional information.

Inadequate	*Adequate*
Accounting Accounting work	Certified public accountant Accountant Accounting machine operator Tax auditor Accounts payable clerk
Adjuster	Brake adjuster Machine adjuster Merchandise complaint adjuster Insurance adjuster
Agent	Freight agent Insurance agent Sales agent Advertising agent Purchasing agent
Analyst	Cement analyst Food analyst Budget analyst Computer systems analyst Procedure analyst Air analyst
Broker	Stock broker Insurance broker Real estate broker Livestock broker
Caretaker or custodian	Janitor Guard Building superintendent Gardener Groundskeeper Property clerk Locker attendant

Appendix A: Examples of Occupation Entries that Need more Detail

Inadequate	*Adequate*
Claims adjuster	Unemployment benefits claims taker Auto insurance adjuster Right-of-way claims agent Merchandise complaint adjuster
Clerk	Stock clerk Shipping clerk Sales clerk or salesperson (person who sold goods in a store)
Consultant	Financial consultant Legal consultant Tax consultant Marketing consultant
Contractor	Construction contractor (specify working or administrative type duties) Managerial contractor Painting contractor (specify administrative, managerial, or working)
Counselor	Educational counselor Personnel counselor Rehabilitation counselor Guidance counselor Marriage counselor
Data Processing	Computer programmer Data typist Keypunch operator Computer operator Coding clerk Card tape converter operator
Doctor	Optometrist Dentist Veterinarian Psychiatrist Chiropractor
Engineer	Civil engineer Locomotive engineer Mechanical engineer Aeronautical engineer Electrical engineer Construction engineer
Equipment operator	Road grader operator Bulldozer operator Trencher operator

Appendix A: Examples of Occupation Entries that Need more Detail

Inadequate	*Adequate*
Factory worker	Assembler (with product listed) Machine operator Forge heater Relief man Turret lathe operator Weaver Loom fixer Knitter, stitcher Punch-press operator Spray painter Riveter
Farmworker	Farmer or sharecropper (person responsible for operation of farm) Farm hand (person who did general farm work for wages; may be a family member) Farm helper (household relative who worked on family farm without pay) Farm manager (person who was hired to manage a farm for someone else) Farm service worker (worker who went from farm to farm to harvest, reap, or do similar operations on contract basis usually using own equipment) Farm supervisor (person hired to supervise a group of farmhands) Fruit picker (person hired to do a particular job) Migratory farmhand (person who moved from place to place to assist in planting and harvesting of crops)
Fireman	Locomotive fireman City fireman (city fire department) Kiln fireman Stationary fireman Fire boss
Foreman (craft or activity involved should be specified)	Carpenter foreman Truck driver foreman Ranch foreman
Heavy equipment operator (type of equipment should be specified)	Clam-shovel operator Derrick operator Monorail crane operator Dragline operator Euclid operator

Appendix A: Examples of Occupation Entries that Need more Detail

Inadequate	*Adequate*
Helper	Baker's helper Carpenter's helper Janitor's helper
Investigator	Insurance claim investigator Income tax investigator Financial examiner Social welfare investigator
Laborer	Construction laborer Laundry laborer Warehouse laborer Oil field laborer
Layout worker	Pattern maker Sheet-metal worker Compositor Commercial artist Structural steelworker Draftsperson Coppersmith
Maintenance worker	Groundskeeper Janitor Carpenter Electrician
Manager	Kitchen manager Office manager Personnel manager Warehouse manager
Mechanic	Automobile mechanic Auto transmission mechanic Airplane engine mechanic Elevator mechanic Copy machine mechanic Auto brake mechanic
Nurse	Registered nurse Nurse-midwife Practical nurse Nurse's aide Student nurse Nurse practitioner

Appendix A: Examples of Occupation Entries that Need more Detail

Inadequate	*Adequate*
Office worker	Typist Secretary Receptionist Computer operator File clerk Bookkeeper Physician's assistant
Program specialist	Program scheduler Data processing systems supervisor Metal-flow coordinator
Programmer	Computer programmer Electronic data programmer Radio or TV program director Production planner
Ranch worker (see Farmworker)	Rancher Ranch hand
Research (field of research should be specified; "associate" or "assistant" should be included if part of the title)	Research physicist Research chemist Research mathematician Research biologist Research associate chemist Assistant research physicist Research associate geologist
Sales worker	Advertising sales Insurance sales Bond sales Driver-sales (route selling) Newspaper sales
Scientist	Political scientist Physicist Sociologist Epidemiologist Oceanographer Soil scientist
Shipping department	Shipping and receiving clerk Crater Order picker Typist Parcel wrapper

Appendix A: Examples of Occupation Entries that Need more Detail

Inadequate	*Adequate*
Supervisor	Typing supervisor Accounting supervisor Shop steward Kitchen supervisor Accounts supervisor Cutting and sewing supervisor Sales director Route supervisor Recreation supervisor Service supervisor
Teacher (occupation for a teacher should be reported at the level taught; subject should be included for those who taught above the elementary level)	Preschool teacher Kindergarten teacher Elementary school teacher High school English teacher College professor (mathematics)
Technician	Medical laboratory technician Dental laboratory technician X-ray technician
Tester	Cement tester Instrument tester Engine tester Battery tester
Works in stockroom, office, etc. (Names of departments or place of work are unsatisfactory)	Shipping clerk Filing clerk Truck loader

Appendix A: Examples of Occupation Entries that Need more Detail

Appendix B

Examples of Industry Entries that Need more Specificity

The following are examples of industries that require special caution in reporting. Included in this listing are examples of entries considered inadequate as well as the correct or adequate listing. Note that the examples of adequate industries do not include all acceptable titles. The most frequently entered inadequate industries appear in bold. Refer to Appendix C for additional information.

Inadequate	*Adequate*
Agency	Collection agency Advertising agency Real estate agency Employment agency Travel agency Insurance agency
Aircraft components, Aircraft parts	Airplane engine parts factory Propeller manufacturing Electronic instruments factory Wholesale aircraft parts
Auto or automobile components, Auto or automobile parts	Auto clutch manufacturing Wholesale auto accessories Auto tire manufacturing Retail sales and installation of mufflers Battery factory
Bakery	Bakery plant (makes and sells to wholesalers, retail stores, restaurants) Wholesale bakery (buys from manufacturer and sells to grocers, restaurants, etc.) Retail bakery (sells only on premises to private individuals)
Box factory	Paper box factory Wooden box factory Metal box factory
City or city government	City street repair department City board of health City board of education

Appendix B: Examples of Industry Entries that Need more Specificity

Inadequate	*Adequate*
Club, private	Golf club Fraternal club Nightclub Residence club
Coal Company	Coal mine Retail coal yard Wholesale coal
County or county government	County recreation department County board of education
Credit company	Credit rating bureau Loan company Credit clothing company
Dairy	Dairy farm Dairy bar Wholesale dairy products Retail dairy products Dairy products manufacturing
Discount house, Discount store	Retail drug store Retail electrical appliances Retail general merchandise Retail clothing store
Electrical parts manufacturing	Electronic tube factory Memory core manufacturing Transistor factory Tape reader manufacturing
Engineering company	Civil engineering consultants General contracting Wholesale hearing equipment Construction machinery factory
Express company	Motor freight Railway express agency Railroad car rental (for Union Tank Car Co., etc.) Armored car service
Factory, mill, or plant	Steel rolling mill Hardware factory Aircraft factory Flour mill Hosiery mill Commercial printing plant Cotton cloth mill

Appendix B: Examples of Industry Entries that Need more Specificity

Inadequate	*Adequate*
Farm	Cattle ranch Crop farm Chicken ranch Cattle and wheat farm Fish farm
Foundry	Iron foundry Brass foundry Aluminum foundry
Freight company	Motor freight Air freight Railway freight Water transportation
Healthcare	Hospital Doctor's office Clinic Private duty
Laundry	Own home laundry (for a person who laundered for pay in own home) Laundering for private family (for person who worked in the home of a private family) Commercial laundry (for person who worked in a steam laundry, hand laundry, or similar establishment)
Lumber company	Sawmill Retail lumberyard Planing mill Logging camp Wholesale lumber
Manufacturer's agent (product sold should be specified)	Jewelry manufacturer's representative Lumber manufacturer's agent Electric appliance manufacturers representative Chemical manufacturer's agent

Appendix B: Examples of Industry Entries that Need more Specificity

Inadequate	*Adequate*
Mine	Coal mine Gold mine Bauxite mine Iron mine Copper mine Lead mine Marble quarry Sand and gravel pit
Nylon factory	Nylon chemical factory (where chemicals are made into fibers) Nylon textile mill (where fibers are made into yarn or woven into cloth) Women's nylon hosiery factory (where yarn is made into hosiery)
Office	Dentist's office Physician's office Insurance office
Oil industry	Oil field drilling Petroleum refinery Retail gasoline station Petroleum pipeline Wholesale oil distributor Retail fuel oil
Packing house	Meat packing plant Fruit canner Fruit packing shed (wholesale packers and shippers)
Pipeline	Natural gas pipeline Gasoline pipeline Petroleum pipeline Pipeline construction
Plastics factory	Plastic materials factory (where plastic materials are made) Plastic products plant (where articles are manufactured from plastic materials)
Public utility (all services should be specified, such as gas and electric utility, or electric and water utility)	Electric light and power utility Gas utility Telephone utility Water supply utility
Railroad	Railroad car factory Diesel railroad repair shop Locomotive manufacturing plant

Appendix B: Examples of Industry Entries that Need more Specificity

Inadequate	*Adequate*
Repair shop	Shoe repair shop Television repair shop Radio repair shop Blacksmith shop Welding shop Auto repair shop Machine repair shop
Research	Permanent-press dresses (product of company for which research was done) Brandeis University (name of university where research was done for its own use) St. Elizabeth's Hospital (name of hospital at which medical research was done for its own use) Commercial research (if research is the main service of the company) Brookings Institution (name of the nonprofit organization)
School (public and private schools, including parochial, must be distinguished, and the highest level of instruction should be identified, such as junior college or senior high school)	City elementary school Private kindergarten Private college
Tailor shop	Dry cleaning shop (provides valet service) Custom tailor shop (makes clothes to customer's order) Men's rental clothing store
Terminal	Bus terminal Railroad terminal Boat terminal Airport terminal
Textile mill	Cotton cloth mill Woolen cloth mill Cotton yarn mill Nylon thread mill

Appendix B: Examples of Industry Entries that Need more Specificity

Inadequate	*Adequate*
Transportation company	Motor trucking Moving and storage Water transportation Air transportation Airline Taxicab service Subway Elevated railway Railroad Car loading service
Water company	Water supply Irrigation systems Water filtration plant
Well	Oil field drilling Oil well drilling Salt well drilling Water well drilling

Appendix C

Examples of Acceptable Entries for Both Occupation and Industry

These are examples of *acceptable entries* of occupation and corresponding business/industry developed by the U.S. Bureau of the Census. They are provided as a guide for proper reporting

Occupation	*Business/Industry*
Attorney	Self-employed
Attorney	Legal aid society
Auditor	Savings and loan
Bookkeeper	Wholesale drugs
Camera operator	Television station
Carpenter	Building construction
Carpet installer	Retail carpet sales and installation company
Cashier	Bank
Chaplain	State prison
Chauffeur	City fire department
Chauffeur	Taxicab company
Chemist	Plastic film manufacturing
Computer programmer	Life insurance company
Delivery driver	Wholesale bakery
Dressmaker	Dressmaking shop
Electrician	Electric light and power company
Field examiner	Veterans Administration (U.S. Government)
Flight engineer	Aircraft company (manufacturing, retail, or wholesale)
Geologist	Petroleum exploration
Insurance agent	Life insurance company
Janitor	City office building
Judge	County court
Mechanic, auto	Engine repair shop
Medical doctor	Board of health (state government)
Miner	Coal mine
Motor operator (retired)	urban transit system

Appendix C: Examples of Acceptable Entries for Both Occupation and Industry

Occupation	*Business/Industry*
Owner (Embalmer and Manager)	Funeral home
Owner/Manager	Retail grocery store
Petroleum analyst	Petroleum refining
Pilot	Commercial airline
Plant manager	Petroleum refinery
President	Business college
Printer (Apprentice)	Printing shop
Production cost estimator	Auto body repair shop
Professor (English)	College
Quarry worker	Marble quarry
Radio operator	College radio station
Registered nurse	Hospital
Senator	U.S. Congress
Shoe designer	Leather footwear factory
Stationary firefighter	Steel mill
Student	Junior college
Supervisor (Weaving)	Cotton cloth mill
Supervisor (Office)	Health and accident insurance company
Teamster (Truck Driver)	Logging camp
Timber cutter	Logging
Tire tester	Tire manufacturing
Weaver	Cotton cloth mill

Appendix D

The Vital Statistics Registration System in the United States

The registration of births, deaths, fetal deaths, and other vital events[3] in the United States is a state and local function. The civil laws of every state provide for a continuous, permanent, and compulsory vital registration system. Each system depends on the conscientious efforts of the physicians, hospital personnel, funeral directors, coroners, and medical examiners in preparing or certifying information needed to complete the original records. For a graphic presentation of the US Vital Statistics registration system, please see Figure I.

Most states are divided geographically into local registration districts or units to facilitate the collection of vital records. A district may be a township, village, town, city, county, or other geographic area or a combination of two or more of these areas. In some states, however, the law provides that records of birth, death, and/or fetal death be sent directly from the reporting source (hospital, physician, or funeral director) to the state vital statistics office. In this system, functions normally performed by a local registration official are assumed by the staff of the state office.

In states with a local registrar system, the local registrar collects the records of events occurring in his or her area and transmits them to the state vital statistics office. The local registrar is required to ensure that a complete certificate is filed for each event occurring in that district. The state vital statistics office inspects each record for promptness of filing, completeness, and accuracy of information; queries for missing or inconsistent information; numbers the records; prepares indexes; processes the records; and stores the documents for permanent safekeeping.

Statistical information from the records is tabulated for use by state and local health departments, other governmental agencies, and various private and voluntary organizations. The data are used to evaluate health problems and to plan programs and services for the public. The National Center for Health Statistics (NCHS), Centers for Disease Control and Prevention, is vested with the authority for administering vital statistics functions at the national level. Data files derived from individual records registered in the state offices – or, in a few cases, copies of the individual records themselves – are transmitted to NCHS.

From these data, monthly, annual, and special statistical reports are prepared for the United States as a whole and for the component parts – cities, counties, states, and regions – by characteristics such as sex, race, and cause of death. The statistics are essential in the fields of social welfare, public health, and demography. They are also used for various administrative purposes, in both business and government. NCHS serves as a focal point, exercising leadership in establishing uniform practices through model laws, standard certificate forms, handbooks, and other instructional materials for the continued improvement of the vital statistics system in the United States.

---

[3] Vital events are defined as live births, deaths, fetal deaths, marriages, divorces, and induced terminations of pregnancy, together with any change in civil status which may occur during an individual's lifetime.

Appendix D: The Vital Statistics Registration System in the United States

Using Death Data for Public Health Action

The National Institute for Occupational Safety and Health (NIOSH) uses death data to evaluate statistical patterns in mortality associated with the usual occupation and industry reported on death certificates (www.cdc.gov/niosh/topics/surveillance/noms/). For example, the link between lung cancer and exposure to asbestos in jobs such as shipbuilding or construction was discovered by evaluating death certificate data. Since 1980, NIOSH together with NCHS and the National Cancer Institute (NCI) have collected and reviewed information from U.S. death certificates de-identified by NCHS and/or the states for public health surveillance and prevention. As one of the few sources of representative data on chronic disease, these data are used to evaluate trends and identify potential risks for acute and chronic disease mortality by industrial and occupational groups.

Lifetime occupation and industry data statements on death certificates can provide information for setting priorities and identifying workplace settings associated with increased or decreased risk of disease and injury. The data have been used in analyses by NIOSH and other researchers to expand knowledge on occupationally-related health outcomes ranging from hepatitis to silica-related systemic silicosis. The information generated is used to inform public health officials about mortality risks by industry and occupation and to identify job-related risk areas.

Because of the reliability of the data for this large and representative, population-based surveillance system, the findings based on this information have been used to support U.S. health and safety recommendations, support prevention activities, and enhance public health education, medical treatment, and screening. Death data are often used for the study of rare disease, workers of minority races and ethnic populations, and workplaces where worker records are not maintained.

NIOSH is charged with conducting occupational hazard and health surveillance to identify trends and assist in setting priorities for research and prevention activities. Since the 1980s, NIOSH has collaborated with NCHS, NCI, the U.S. Census Bureau and other federal agencies, and state health departments to capture and code usual occupation and kind of business or industry on death records for surveillance of occupationally related diseases, injuries, and exposures. For information about the classification systems used to code the industry and occupation text that funeral directors and their staff record, see the U.S. Census Bureau's Industry and Occupation Overview page at http://www.census.gov/hhes/www/ioindex/overview.html.

U.S. states that use the 2003 revised standard death certificate and enter electronically the text narrative of occupation and industry can now code these data for statistical use using the new NIOSH desktop computer-assisted or autocoder coding system (http://www.cdc.gov/niosh/topics/coding/software.html). An internet-based version of the coding system will become available in 2013. NIOSH can also help states improve their coding of usual occupation and type of business or industry items on the death certificate by conducting at least two training classes per year in the Census 2000 codes that are now being used in conjunction with the revised death certificate (http://www.cdc.gov/niosh/topics/coding/training.html). NIOSH has responsibility for the quality of occupation and industry narratives and their coding, and conducts quality control in collaboration with NCHS on the coded occupation and industry data for states.

Appendix D: The Vital Statistics Registration System in the United States

Figure 1: The U.S. Vital Statistics Registration System with Roles of State and National Agencies

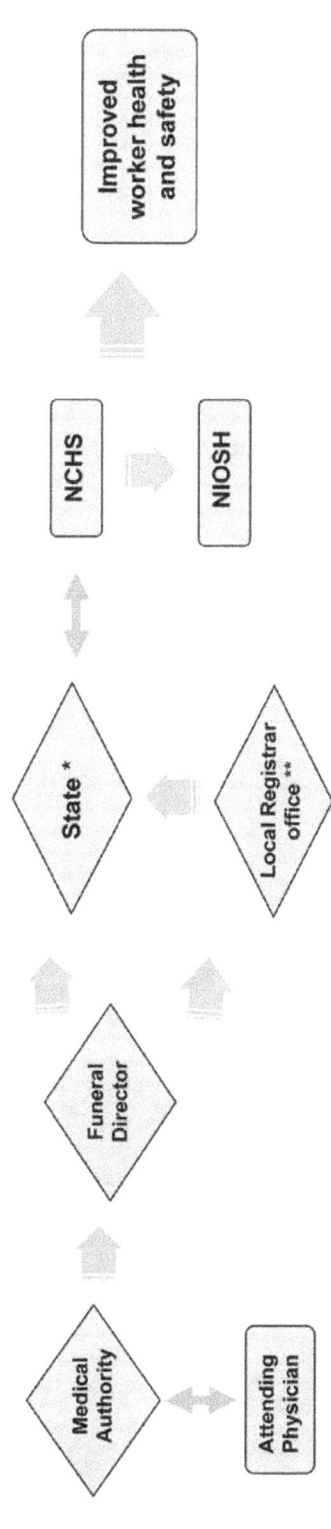

The Roles

Hospital/Physician

When death occurs in hospital, may initiate preparation of certificate:

1. Completes information on name, date, and place of death;
2. Obtains certification of cause of death from physician; and
3. Gives certificate to funeral director.

Funeral Director

1. Obtains personal facts about decedent, including usual occupation and corresponding industry, and completes death certificate.
2. Obtains certification of cause of death from attending physician or medical examiner or coroner.
3. Obtains authorization for final disposition per state law.
4. Files certificate with local office or state office per state law.

State Registrar

1. Queries incomplete or inconsistent information.
2. Maintains files for permanent reference and is the source of certified copies.
3. Develops vital statistics for use in planning, evaluating, and administering state and local health activities and for research studies.
4. Compiles health-related statistics for state and civil divisions.
5. Sends data derived from records or copies of records to the National Center for Health Statistics and the National Institute for Occupational Safety and Health.

Federal Government Agencies:

National Center for Health Statistics (NCHS)

1. Prepares and publishes national statistics of births, deaths, and fetal deaths; constructs the official U.S. life tables and related actuarial tables.
2. Conducts health and social research studies based on vital records.
3. Conducts research and methodological studies on vital statistics methods.
4. Maintains a continuing technical assistance program to improve the quality and usefulness of vital statistics.

National Institute for Occupational Safety and Health (NIOSH)

1. Provides training of Industry and Occupation (I&O) coders, quality assurance, and provides assistance for coding I&O to state programs.
2. Addresses current issues in workers' health and safety through action.
3. Prioritizes and conducts occupational safety and health research studies; improves policy and recommends workplace regulations.
4. Augments occupational physicians' knowledge.

Footnotes:
* May be local registrar or city or county health department.
** Some states do not have local vital registration offices. In these states, the certificates or reports are transmitted directly to the state registrar office.

www.ingramcontent.com/pod-product-compliance
Lightning Source LLC
Chambersburg PA
CBHW081800170526
45167CB00008B/3260